M. M. Whelan

Hand-Book of the Boston and Hingham Steamboat Company

Vol. 1

M. M. Whelan

Hand-Book of the Boston and Hingham Steamboat Company
Vol. 1

ISBN/EAN: 9783337408787

Printed in Europe, USA, Canada, Australia, Japan

Cover: Foto ©berggeist007 / pixelio.de

More available books at **www.hansebooks.com**

HAND-BOOK

OF THE

BOSTON AND HINGHAM

STEAMBOAT COMPANY.

By M. M. WHELAN.

A COMPLETE GUIDE TO THE

GREAT WATERING PLACES OF BOSTON.

HOW TO REACH AND ENJOY

NANTASKET BEACH,
"THE CONEY ISLAND OF BOSTON,"

MELVILLE GARDEN,
"THE ROCKY POINT OF BOSTON,"

DOWNER LANDING,

HINGHAM, AND HULL.

COMPLETE NARRATIONS OF THEIR

VARIOUS ATTRACTIONS, HOTELS AND BATHING HOUSES,
AND SCENES IN THE HARBOR,

AND

CORRECTED TIME-TABLE AND RATES OF FARE.

Only Authorized Guide-Book to Boston's Popular Seaside Resorts.

Illustrated, and containing a Correct Map of the Harbor.

PRICE, 10 CENTS.

BOSTON:
ALFRED MUDGE & SON, PRINTERS, 34 SCHOOL STREET.
1880.

INTRODUCTORY.

THE first soft wind that blows from the south in the early spring comes laden with the promise of hope to the weary workers in the Lord's vineyard. It whispers of the merry songs of the birds, of verdure, and the sweet perfume of the rose. It breathes inspiration into the soul of toiling humanity, and tells the heart with gentle voice that new life will soon adorn the field all brown and bare, and the cold, dreary months of winter will soon pass away and the joys of summer days be again renewed.

The days pass on, and the slow action of nature brings about the awakening of the promised transformation; and as the buds appear and the vesture of earth puts on its freshest color, the heart grows more buoyant. The thoughts of the rich and poor alike begin to evolve plans for their summer enjoyment. Soon will the tedious care of the workshop and the counting-room be thrown off, and the din of the narrow city's fens be changed for the bracing atmosphere of old ocean and the balmy air of the country seat.

The merchant, having the means to gratify his every desire, can easily prepare his schemes for recreation; but the middle classes and the ever-present poor who seek for freedom must find their source of joy in those opportunities which come within a radius of their income. Fortunate, indeed, it is for them that there is an Elysium, and that there are men with generosity of soul large enough to take in the desires of the masses

Vincent, Hathaway & Co.

and give to them the means by which they can be transported to these fields of pleasure.

The eager anticipation of the excursionist has been fully met by the more anticipating benefactors, and to-day we have the Boston and Hingham Steamboat Company, who have done more good in the past toward the morals and health of the swarming population of Boston during the summer than any other source of influence; and to this company belongs the credit of the thousand and one improvements that have been made at the various points of interest in our harbor from time to time. The company own the best steamers in every respect that ply their wheels in our beautiful harbor; and possessing every known facility for comfort, safety, and elegance, their popularity and patronage has become simply immense. In view of a still greater amount of travelling over their line this year than ever before, the company, in order to gratify the every desire of their patrons, have published this Guide, which will be found to be interesting, and explains to those unfamiliar with our harbor the many points of interest, as well as other pleasant features.

PUBLISHER'S ACKNOWLEDGMENT.

The Publisher is indebted to the following gentlemen for many essential favors in assisting in the compilation of this Hand-Book : —

      Mr. GEORGE LINCOLN, of Hingham.

      Mr. FRANK T. ROBINSON, of Boston.

      Mr. MARLTON DOWNING.

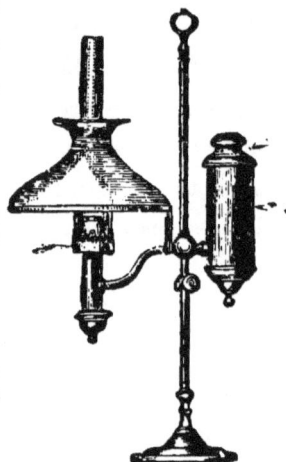

The Boston and Hingham Steamboat Company.

BRIEF HISTORY OF THIS POPULAR LINE.

THERE can be no better introduction to this guide-book than a slight outline sketch of the history of the Boston and Hingham Steamboat Company and the boats which have from time to time been used by them.

In the year 1818, the "Eagle," under the command of Capt. Clark, made the first steamboat trip from Boston to the town of Hingham. She made irregular trips during that season, and it was not until 1819, one year later, that she was put on the line to make daily passages to and from these places. She was then under the charge of Capt. Barnabas Lincoln. She was a small boat, compared with those of to-day, accommodating two hundred passengers. She ran until 1821, after which, until the

year 1829, we have no account of any boat except the packets plying between Boston and Hingham. A very singular incident is that during the year 1822 there is not a single notice of

a steamer in Boston Harbor; where they all went to nobody knows. The "Lafayette" was the next boat. She began run-

ning in 1829, commanded by Capt. George Thaxter, who run her until 1830, when he was relieved by Capt. George Beal, of Cohasset. She was much smaller than her predecessor, the "Eagle," and even in those primitive days of steamboating was not very highly esteemed. Before she was purchased by the Hingham people, she was called the "Hamilton," which name always remained on her stern. There is an account of her

being caught in a squall off the Castle, on her trip to Boston, and being obliged to put back to Hingham The "Gen. Lincoln" followed the "Lafayette." She made her first trip June 16, 1832, under Capt. George Beal. It was at this time that the present company was formed. The "Gen. Lincoln" was built at Philadelphia; measured 96 feet long, with 22 feet beam; side-wheels, 14 feet in diameter. She had 2 boilers, 2 walking-

beams, burnt wood, and was run under about 20 pounds steam. The "Mayflower," built in New York expressly for this com-

pany, was put upon this route about the first of July, 1845.
She had numerous commanders. Capt. George Beal was her
first, and Capt. Rowell her last in this line. In 1856 she was
bought by New York parties. The "Mayflower" was 133 foot
8 inches in length by 42 feet 2 inches beam; 8 feet 8 inches in
draught; 262 and 69-95 tons burden, old measurement. Her
wheels were 22 feet diameter, 30-inch cylinder with 10-foot
stroke, run under 30 pounds steam, and accommodated 800 passengers. The fifth boat was the "Nantasket," Capt. A. L.

Rowell. She was built in New York in 1857, and placed upon
this line in 1858. Her original proportions were: length, 146
feet 6 inches; breadth, 25 feet 4 inches; depth, 8 feet 2 inches;
wheels, 24 feet in diameter; and 285 and 89-95 tons, old measurement. She was considered the fastest boat in the harbor at that time. In 1862 she was in government employ,
South. She was relieved during that summer by the "Gilpin"
and "Halifax," the latter a stern-wheeler. In 1863 the "Nantasket" returned to Boston, and underwent numerous altera-

tions, besides changing her name to the "Emeline." The
"Rose Standish," now running, was built in Brooklyn, N. Y.,
in 1863. She is 392 and 93-100 tons, new measurement. Capt.
Jones was commander in 1864, Capt. Brown in 1866, Capt. Good
in 1867, and she is now commanded by Capt. Bird. The "John
Romer" was also built at Keyport, in 1863, and was intended
to run between New York and Greenwich, Conn. It was purchased by the Boston and Hingham Steamboat Company, and

placed upon this line in 1866. Capt. Good commanded three seasons, as did Capt. Collins. It is now under the command of Capt. Wm. H. Sampson. The "Governor Andrew" was built

at Green Point, N. Y., in 1874, by the same parties who built the stanch and well-tried steamer "Rose Standish," Messrs. Folkes & Lawrence. Her length is 159 feet, with 27 feet beam, and depth, 9 feet. Her measurement is 503 and 1-100 tons, and

she is commanded by Capt. George F. Brown. The elegant new steamer "Nantasket" was launched May 15, 1878, from the yard of Messrs. Pierce & Montgomery, Chelsea. She is 173 feet long by 29 feet breadth of beam, and 9 feet in depth.

Her tonnage is 498 and 23-100. Capt. Chas. E. Good has charge of her.

. . . .

What a delightful sensation steals over the mind of the excursionist as he sits on the deck of the steamer before leaving the pier! The day is his, and he already abandons himself to the

most pleasurable thoughts. The smell of the salt sea air invigorates his being, and in anticipation he gazes seaward with eyes all anxious to view the broad ocean and the white sand beach.

> " Be it the summer noon, a sandy space
> The ebbing tide has left upon its place;
> Then, just the hot and stony beach above,
> Light twinkling streams in bright confusion move
> (For, heated thus, the warmer air ascends,
> And with the cooler in its fall contends)."

Soon the vibration of the great engines is heard, and with a gentle movement the good steamer leaves the wharf, and then a most delightful series of sights greets the eye.

WINTHROP ISLAND.

The view as we head toward the harbor is a pleasing one. There are the pleasure yachts and carrying vessels, the busy tug-boats straining away with a big ship behind them, and far away in the middle distance and horizon are the oval crests of the islands that here and there dot the water's surface. After passing the broad opening, the first island that is plainly seen, or enough so to distinguish its character and formation, is Governor's, or, as it is now called, Winthrop Island. It derives its name from Gov. Winthrop, to whom it was given at a very early period in the history of Boston by the colonial Legislature. This noted island first took its name from Roger Conant, a distinguished early settler of Plymouth. The first known of this island is that on July 2, 1631, it was "appropriated to publique benefits and uses." From papers now in possession of the New England Historical Genealogical Society, we copy the following: On the 3d of April, 1632, at a Court of Assistants, " the island called Conant's Island, with all the liberties & privileges of fishing and fowling, was demised to John Winthrop, Esq., the psent gounr., and it was further agreed that the said John Winthrop did covenant and promise to plant a vineyard and an orchard in the same, and that the heirs or assigns of the said John Winthrop for one & twenty yeares payeing yearely to the gounr, the fifth parts of all such fruits & proffits as shalbe yearly raysed out of the same, and the lease to be renewed from time to time onto the heirs and assigns of the said John Winthrop, & the name of the said ileland was changed & is to be called the Gounr's Garden."

The island continued in the possession of the Winthrop family

A LARGE LINE

— OF —

FANCY

STRAW MATTINGS

WHICH WE SELL AT

25 CENTS

PER YARD.

CHIPMAN'S SONS & CO.,

Cor. Court & Hanover Sts.,

BOSTON.

till 1808, when they sold a portion of it to the government, for the purpose of erecting a fort thereon. This, when built, was called Fort Warren, in honor of Gen. Joseph Warren. Since then another fort has been erected on George's Island, and this name was transferred to it. A new fortification is now in process of erection on the summit of the high hill on Governor's Island, and has been named Fort Winthrop, in remembrance of

FORT WINTHROP.

the first governor, to whom it was granted. This fort is of great strength, and has a very commanding position. Its batteries are nearly all underground, and connected with the citadel (the top of which can be seen at the highest part of the island) by underground passages; and the water battery that will be observed on the southerly side of the island is of great advantage to the defence, controlling, as it does, a large extent of flats, which are very shoal except at the highest tides

FORT INDEPENDENCE.

The next island which is passed, and lying to the south of the steamer's course, is Castle Island. It is situated almost directly opposite Fort Winthrop, and can be easily recognized by the granite fortress and earthworks placed thereon. It is considered one of the most prominent forts in Boston Harbor. In 1634 the idea was first conceived of erecting a fort upon Castle Island. Capt. Edward Johnson, of Woburn, thus speaks of the fort, in a work published in 1654: " The Castle is built on the

northeast of the Island, upon a rising hill, very advantageous
to make many shot at such ships as shall offer to enter the harbor without their good leave and liking; the Commander of it

FORT INDEPENDENCE.

is one Captain Davenport, a man approved for his faithfulness
and skill. The master cannoneer is an active engineer; also the
castle hath cost about four thousand pounds, yet are not this poor
pilgrim people weary of maintaining it in good repair. It is
of very good use to awe any insolent persons, that putting con-
fidence in their ships and sails, shall offer any injury to the people or contemn their governments, and they have certain signals
of alarums which suddenly spread through the whole country."

When the British evacuated Boston they destroyed Castle
William, as it was then called; and after the Provincial forces
took possession they repaired it, and its name was changed to
Fort Independence in 1797, President John Adams being present on the occasion. This island was noted for years as a duel-
ling ground. On the glacis of the fort is now standing a memo-
rial of one of these unfortunate affairs, on which is the following inscription: —

NEAR THIS SPOT

ON THE 25TH DEC. 1817

FELL

LIEU. ROBERT F. MASSIE

AGED 21 YEARS.

The castle was used as a place of confinement for thieves and other convicts sentenced to hard labor, from 1785 till the State Prison was built in 1805, at Charlestown. Within a few years a substantial stone fort has been erected in place of old Castle William. A number of prisoners were confined here during the war of the Rebellion, and several deserters were executed.

SPECTACLE ISLAND.

Spectacle, or, as it is sometimes called, Ward's Island, on account of Ward's rendering works now placed there. In 1634 it was rented to the city of Boston for one shilling. In 1717 this island was sold to the Province for the sum of £100, in bills of credit, for the purpose of erecting a " Pest House for the reception and entertainment of sick persons coming from beyond the sea, and in order to prevent the spreading of infection."

LONG ISLAND LIGHT.

Things must have progressed slowly at the island; for in 1720 we find it was voted " that the selectmen of the town of Boston be desired to take care for the furnishing of the Public Hospital on Spectacle Island, so as to make it warm and comfortable for the entertainment of the sick." In 1736 the hospital was removed to Rainsford's Island, and Spectacle Island ceased to be of any great importance.

LONG ISLAND.

That long stretch of land on your right is Long Island, and is about three quarters of a mile below Spectacle. It used to

THOMPSON'S ISLAND.

be thickly covered with a timber growth, which was as far back as 1640. There have been various owners; and in 1847 the Long Island Company bought all the island, with the exception of the East Head, built a substantial wharf, and erected the Long Island House, laid out streets and cut up the land into lots for building, starting a real-estate speculation on the island; which, however, was not successful, as but a few buildings were erected.

The lighthouse on the East Head was erected in 1819. Its tower is twenty-two feet in height, built of iron and painted white, with a black lantern, containing nine burners, and is about eighty feet above the level of the sea, with a fixed white light that can be seen on a clear night about fifteen miles. It is situated in a square enclosure of ground on the summit of the Head. Within the square is a comfortable stone house for the keeper.

The northeastern part of the Head has for years been gradually washing away, and it was not until recently that the government built a sea wall for its protection from the inroads of the surf.

DEER ISLAND.

Deer Island lies directly north of the East Head of Long Island, and between these two islands the boats pass. It is now used by the city, and the House of Industry has been erected thereon, which was removed from South Boston (where it was formerly situated) in 1848. The House of Reformation and Almshouse were removed from the same place in 1858. The large brick building which is so prominent was built in 1850.

NIX'S MATE.

Nix's Mate is one of the greatest points of interest in the harbor. It is a tall pyramid placed upon a square stone base.

NIX'S MATE.

The beacon as it stands, from base to peak, is thirty-two feet in height. It rests upon a shoal, which, at low tide, shows plainly out of water. There are two very interesting stories connected with its origin; viz., one is that the mate of a certain captain by the name of Nix was executed upon the place for killing his

ALFRED MUDGE & SON,

Legal, Mercantile and Commercial

PRINTERS.

Our Establishment is one of the largest and oldest in New England, and its facilities for the execution of Printing of every description are unsurpassed.

Send for estimate.   We guarantee all work to give satisfaction.

34 School Street,

BOSTON, MASS.

master. The other is that Nix had been at one time connected with some piratical enterprise; his mate being caught was here doomed to death. In either case the mate protested his innocence, and in proof thereof he asserted that in a certain number of years the island would be entirely washed away. Whether the man was innocent or guilty, of course nobody knows; but it is positively certain that the once beautiful little island has quite passed away from sight, and the black object now upon it is a fitting pile to commemorate the memories of the spot.

GALLOP'S ISLAND.

Bearing southeast from Nix's Mate, is Gallop's Island, so called as it was once the property of John Gallop, a harbor pilot. It forms the southern border of what is termed "The Narrows," the channel through which the boats usually pass, though sometimes they go to the southward.

This property came into possession of the city of Boston in 1860, purchased from Charles Newcomb for the sum of $6,500.

LOVELL'S ISLAND.

The island lying to the northward of the Narrows is called Lovell's Island. It has been the scene of numerous shipwrecks; the most important of which was the loss of the French frigate "Magnifique," the flag-ship of the French Squadron, under the command of Count D'Estaing, in 1782. The vessel struck on Man-o'-war Bar, which extends from the extreme westerly end of the island. The loss of this ship (which it was said was owing to the carelessness of the pilot) was a very serious matter to the Americans, as the French were the allies of Young America in her Revolutionary struggles.

GEORGE'S ISLAND.

Next in our course lies George's Island, on which Fort Warren is situated. The steamer leaves it on the port (or left-hand) side as she swings out of the Narrows and heads for Hull. It was sold to the city of Boston, in 1825, by Caleb Rice, together with Lovell's Island, for $6,000, and afterward transferred to the United States. There is a strong sea-wall nearly surrounding the island. Fort Warren was built upon it in 1850, and is considered the strongest of our harbor fortifications. This fort was used for rendezvous purposes during the

late war. It is also noted as being the prison of Mason and
Slidell, two Confederate commissioners sent by their govern-

FORT WARREN, BOSTON HARBOR.

ment to represent the Southern Confederacy in France and
England.

KEEP COOL AND COMFORTABLE

DURING THE

HEATED TERM

BY A VISIT

DOWN THE HARBOR

IN THE BOATS OF THE

Hingham Steamboat Co.

BUG LIGHT.

You will see, if looking straight out over the water in a northeasterly direction from Fort Warren, a house of octagon shape, and apparently built on spiles. It was built in 1856, is painted red, and its beacon light is seen seven miles away, the lantern being thirty-five feet above the level of the sea.

BOSTON LIGHT.

That white-body and black-top lighthouse away to the left middle distance is the well-known Boston Light.

The first lighthouse was built in 1715. It was much injured by fire in 1751, and was struck several times by lightning. During the Revolutionary war it fared hard. The present lighthouse was erected in 1783, but has been several times refitted since then with improved apparatus; and in 1860 the

BOSTON LIGHT.

old tower was raised in height, it now measuring ninety-eight feet above sea level. The white tower, with its black lantern and revolving light, can be seen at a distance of sixteen nautical miles, if the weather be fair and the sky clear, and is an imposing object when viewed from vessels on entering or leaving the harbor.

ENGELHARDT'S,

173 and 174 Tremont Street,

FACING THE COMMON.

The Best Appointed and Most Elegant Establishment in Boston.

THIS superb Restaurant, Café and Ice Cream Saloon, occupying four
stories of a large and beautiful building, though but recently opened,
has attained a popularity beyond precedent.

(1.) Upon the first floor is a Café and Dining Room for Gentlemen.
A daily bill of fare, comprising the best the market affords.

(2.) The second floor is occupied by an elegant Restaurant for Ladies.
A choice *menu;* also the most delicious Ice Creams, Sherbets, Cake and
Coffee.

(3) and (4.) On the upper floors are Private Dining and Supper
Rooms.

The view from the front windows at all hours is an unfailing charm.
The kitchen is in charge of the most experienced cooks. The finest
Wines of all vintages are served in perfect condition. The famous
TIVOLI and ROCHESTER Lager Beer always on draught. Finest Cigars,
Liquors and Cordials.

With all these advantages, and with scrupulous neatness and good
taste that pervade the establishment, visitors may be sure of having
every reasonable want satisfied.

ENGELHARDT & CO.

ARE AGENTS FOR THE

TIVOLI AND ROCHESTER BEER,

And offer either kind in Bulk or in Bottles.

OFFICE IN THE REAR OF THE MAIN BUILDING, AT

RAINSFORD'S ISLAND.

Rainsford's, or, as it is often called, Quarantine Island, lies southeast of Fort Warren, distant about two miles and a half.

VIEW OF HULL.

It has long been used by the city as a place of detention for patients afflicted with contagious diseases; but at the present time it is occupied mostly by paupers. It became the property of the colony, in 1736, for the sum of £570.

HULL.

That projecting, flat point of land on the left which you are now approaching is Hull Point; and as you pass through the "Gut," with Pettrick's Island on the right, you will notice a strong and often terrific current, which is caused by the shallow waters and sand-bars. On the left, in a sort of bog-like opening, the pier is seen jutting out from the land; while away over the first rising ground is visible a tower and pole, which is called Telegraph Hill.

Hull is not, as many would suppose, an island, but the extreme end of the mainland. It is an incorporated town, and Nantasket lies within its limits. It was one of the places first visited by harbor excursionists, and it has been for years considered a popular place for summer visitors. It is also a favorite resort for yachtmen, as here they find a good harbor, and can have a pleasant time ashore. The hotel, Oregon House, is pleasantly located near the landing.

PETTRICK'S ISLAND.

This island is about a mile long and a quarter wide, and is quite hilly. It is excellent pasturage ground, and during the hot weather is used by the city Bohemians as a camping-ground.

SHEEP'S ISLAND.

Now comes a stretch across the pleasant bay, with Sheep's Island a little to the southward. It is a small, low island of no note and little value. Farther on, the steamer leaves Bumpkin Island on its port. This is a very small island, owned by Harvard College, having been bequeathed to that institution by Samuel Ward, its former owner.

WEIR RIVER.

We have now reached the month of the Weir River, with White Head on the port and World's End on the starboard side. Here is a scenic effect that is truly marvellous, and reminds the traveller of the Palisades opposite Hastings-upon-Hudson. Though not four hundred feet in height, still the picturesqueness of the shore, with the undulating lines of rocks and earth and the harmonious contrast of the water reflecting the sky and meadows as we pass near the shore, is remindfully impressive. Here are to be seen, quietly grazing on hillside and vale, sheep, horses, and cows, and the sight is always a pleasant one to those unaccustomed to such views. There is not a prettier spot in Boston Harbor, nor one better known or more frequented by yachtmen. Scarcely a day passes in summer that there cannot be seen the masts of numerous yachts lying here at anchor, or sailing up the river. Here is *the* acknowledged camping-ground *par excellence*. Fresh water can be had in abundance; berries, in their season, are very plentiful. You will notice, as the boat proceeds up the river to her pier at Nantasket, the very circuitous and crooked route she is obliged to make. At low tide one unacquainted with the river would almost fear of her striking the rocks as she swings round the bend after passing World's End. But it is necessary for her to keep near the south shore, as it is only here that there can be found water enough to float the steamer. After numerous sharp turns and a few cable-lengths of straight course, the boat hauls alongside of the pier at Nantasket, which has been enlarged since last season, being now upward of one hundred feet wide at any point. Here the passengers will disembark, and enjoy the "Coney Island of New England."

NANTASKET BEACH.

"Pleased I look back and forward
And view the tranquil tide
That laves the pebbled shore."

After a short stroll on the plank walk which reaches to the beach, the pleasure seeker finds himself viewing the widest possible range of Massachusetts Bay. Away to the right is Cohasset, and on the left middle distance Point Allerton is seen. One vast stretch of ocean, unbroken save by the great ships and steamers that ply their pathless course to and fro, greets the eye as you gaze toward the far-off horizon. You are invigorated with the sight, put into genial and happy temper; the horizon opens, and you are full of good-will and gratitude to the

Rockland Café,

DIRECTLY AT THE HEAD OF THE STEAMBOAT LANDING.

The Best Equipped Establishment of its kind on the Beach.

DANCING, BOWLING, BATHING, SWINGS, ETC.

Fish Dinners a Specialty,

AT THE LOWEST PRICES.   .

LARGE OR SMALL PARTIES ACCOMMODATED

At the Shortest Notice.

SEPARATE VERANDAS.

No place at Nantasket can off r the Attractions which the

Cause of causes. Now the poor and the rich are alike, and the sentiment of the poor woman can here be appreciated, who, coming from a wretched garret in an inland manufacturing town for the first time to the seashore, gazing at the ocean said, "She was glad for once in her life to see something which there was enough of."

COHASSET ROCKS.

Here is an inexhaustible wealth of the great truths of nature. There is music in the sound of the rolling surf as it washes up on the firm, sandy beach. Up and down the many miles of shining shore you wander, and gather the curious relics tossed up from the

"Deep, dark, and unfathomable caves of ocean."

After a half-day's roaming, the body being weary, the mind is attracted to the many glad pleasures offered by the various houses and cafés. There is the bathing house, the boating facilities, bowling, billiards, dancing, and everything that an enterprising summer resort can offer.

Nantasket Beach has been a place of summer resort more or less for forty or fifty years. The first hotel on its shores was Warwick's Tavern, now Mr. Arthur Pickering's residence, on the road as you go towards the telegraph office. It is a brown house with a long shed attached; and in days of yore was the scene of many a brilliant ball, while its patrons included some of the finest representative people of the city of Boston. Persons of high social standing from adjoining towns used to visit there. Gradually the now famous beach became more generally known, and it became the residence during the heated term of many of the wealthiest people of the Hub. A great many people hired or owned cottages somewhere along its cool and pleasant shores. In 1845 the first attempt was made at building a hotel of any size, when Mr. David Whiton, at the head of a company, which included Mr. N. Ripley, projected and carried out the idea of building the famous Rockland House. Mr. Ripley was the first proprietor of the Rockland, which then contained about forty rooms. The pier was built in 1869. Since that time there have been marvellous improvements everywhere upon the beach; noticeably that upon the Jerusalem Road, which is lined with the finest class of summer residences imaginable. Many of the barren pastures which were once to be seen have been metamorphosed and beautified by the improvements that Boston capital has brought about. Before the present substantial pier was built by the Boston and Hingham Steamboat Company, an attempt was made about ten or twelve years prior to enlarge the Rockland and build a pier, but it was given up as impracticable at that time. When this steamboat company ran their first boat, the "Rose Standish," they had all the passengers they could carry, and business steadily increased until they have had to enlarge their capacity to four boats. The travel has been so large that it has had a tendency to lengthen the seasons. They started their first boat not far from the first of July, considering that there was not business enough to warrant them in starting earlier. Now all through the month of May they have carried people down there. Taking into consideration that all these hills are covered with occupied cottages, besides the increased travel, it is safe to estimate that there is not less than a steady population of upwards of 10,000 people. All this has been brought about within the last decade. When one considers the picturesque features of this section, the accessibility to rural scenes, which remind the visitor of the Granite Hills of New Hampshire and the towering peaks of the White Mountains, the delightful and invigorating sea breezes, the absence of malaria arising from any salt marshes, the dry and healthful soil, the superb sanitary arrangements, the elevated social standing of the inhabitants, the certificates of renowned physicians that nowhere else in the United States can be found such an admirable place for invalids: when one considers all these things, he cannot but acknowledge that Long Branch is distanced, and Nantasket must be called the equal, if not the superior, of Newport in all that can be desired at a seaside resort.

HOTEL NANTASKET.

The Hotel Nantasket, that sits upon the crest of the beach opposite the steamboat landing, is one of the most elegant summer hotels along the American shore. Messrs. Hall and Whippie, the well-known proprietors of Young's Hotel, Boston, have charge of Hotel Nantasket, and will do their best to sustain the reputation which they have hitherto possessed. To meet the demands of their increasing patronage, they have been obliged to enlarge their accommodations, and have now under their control an establishment whose frontage measures five hundred and twenty feet along the beach. The hotel contains a number of apartments fitted up as club and private dining rooms. In fact, there is everything found at this hotel to satisfy the most fastidious of pleasure seekers. The new and elegant veranda extending a thousand feet (connecting the hotel with the Rockland Café) offers its cool retreat to those who wish protection from the rays of the sun.

Messrs. Hall & Whipple have made wise selections in choosing for their officials gentlemen who understand their business, and who have had years of experience; among whom is Mr. W. A. Russell, he being generally in charge. Detective A. P. Dearborn has charge of the Nantasket Beach Hotel Company's property; and as Mr. Dearborn is well known, it is not likely that there will be any disturbance from offenders who may happen on the grounds. Then, too, the guests need fear no annoyance either from ruffians or fire, Mr. Dearborn having a perfect system of drill and other arrangements which will most effectually dispel any alarm.

The Cadet Band will be stationed at the hotel during the summer, and give their daily and evening concerts. The 9.30

The New and Very Large

HOTEL NANTASKET.

ON THE EUROPEAN PLAN.

This elegant establishment, entirely new last year, fully justifies the utmost anticipations of its projectors. Its success was assured from its opening, and its popularity steadily increased to the close of the season, and since that time its immense proportions have been increased by the addition of

SIXTY NEW ROOMS,

Including several dining-rooms for large or small parties.

Situated on Nantasket Beach, near the landing of the Boston & Hingham Steamboat Co., it affords the citizens of Boston and vicinity a resort easy of access, where they may pass a few hours or a day of delightful and comfortable recreation during the heated term.

The Hotel will be strictly first-class in all its appointments, and the cuisine will be characterized by the same excellence which has made Young's Hotel so popular.

The steamboats, which will run nearly every hour during the day and evening, will enable persons from the city to leave town after business hours, secure an ocean bath and a nice dinner or supper, and return to the city refreshed and invigorated.

DURING THE SEASON,

Day and Evening Concerts

WILL BE GIVEN BY THE

BOSTON CADET BAND.

WALTER EMERSON, Cornet Soloist,
            J. THOMAS BALDWIN, Musical Director.

ELECTRIC LIGHTS WILL ILLUMINATE THE HOTEL AND SURROUNDINGS EVERY NIGHT.

HALL & WHIPPLE,

(OF YOUNG'S HOTEL,)

PROPRIETORS.

p. m. boats will afford a most excellent opportunity to lovers of the beautiful to view the ocean by night, which will be enhanced by the electric light and the dulcet strain of music.

THE ROCKLAND CAFÉ,

one of the most popular places of its kind on the beach, ranks equal to its neighbor, the Nantasket, in its culinary department· It is at the present time, as well as for a number of seasons past, under the proprietorship of Mr. H. Ripley, a gentleman thoroughly capable of discharging the arduous duties which devolve upon him in a most acceptable manner. Mr. J. Mc-Laughlin is manager of the Café. Connected with it is a large dancing hall and a picnic pavilion, arranged into compartments so as to accommodate large or small parties.

Surf bathing can be enjoyed on the beach with perfect safety, at the finest bath-houses in the country, where bathing-dresses are always kept to let.

A little to the southward, and standing upon the brow of Atlantic Hill, is placed

THE ATLANTIC HOUSE.

Its site is one of the finest in New England, commanding as it does, from its verandas, an unbroken view of the Massachusetts shore from Marblehead to Scituate. This hotel is one of the largest at Nantasket, capable of accommodating one hundred and seventy-five regular boarders, while their kitchens are under such excellent control that they can readily supply their tables for two hundred and fifty transient visitors. This establishment is under the management of Mr. I. L. Damon, who has had charge of it for a number of years.

THE ROCKLAND HOUSE,

Fully as well equipped for the accommodation of its patrons as any of those we have mentioned, is kept by Mr. J. S. Doyle, the popular proprietor of the St. James, Boston. It is situated upon Atlantic Hill, and lacks but little of the fine view accorded to its neighbor, the Atlantic.

Crossing the hill, and descending the southern slope towards Stony Beach, we meet with numerous smaller hotels, the most prominent of which is the

PACIFIC HOUSE.

It is not much smaller, however, than the Atlantic or Rockland hotels, as can be seen by its capacity for accommodating

NEW PACIFIC HOUSE.

Opened June 1, 1878.

NANTASKET BEACH, NANTASKET, MASS.

W. B. HATHAWAY, Proprietor.

———— •‑• ————

This house was erected two seasons ago on the site of the old one, and is newly furnished throughout. Its

Broad Piazzas, Extensive Halls and Well-Ventilated Apartments

insure to its guests an unusually pleasant and comfortable summer home. Its location is unsurpassed by any house on the shore, commanding an unbroken view of old ocean, and furnishes superior advantages for

BOATING, BATHING, FISHING AND GUNNING,

making one of the most pleasant and romantic places on Massachusetts Bay. A few rods in rear of house, a

LAKE TWO MILES IN LENGTH

affords a safe and pleasant place for ladies and children to row, boats being always in readiness for that purpose. The hotel is

Easily Accessible by Boat or Rail,

being only fifty minutes' sail from Boston by steamer.

one hundred and fifty guests. Mr. William B. Hathaway has charge of the house. The scenery upon this side of the hill offers a marked contrast to that presented by the northern slope. Here it is rugged and rough, very bold and rocky, forming a much grander picture than the smooth, unbroken surface of the beach at Nantasket. The view from the piazzas of the Pacific House is truly magnificent, and Mr. Hathaway, the genial proprietor, cordially invites the visitors to avail themselves of the cool shade afforded by his veranda.

Leaving the hill and returning to the beach, we find, besides the hotels which we have mentioned, numerous cafés and less pretentious establishments, which tend to increase the material comforts of summer visitors. The whole of this magnificent stretch of beach is most beautifully and perfectly illumined nightly with very powerful electric lights. The new broad-gauge railroad, which extends from the steamboat landing to Point Allerton, the extreme northern end of the beach, is a great accommodation, inasmuch as it does away with the slow and tedious ride in the barges, which has hitherto been the only means of conveyance from point to point.

ROSE STANDISH HOUSE.

DOWNER LANDING.

Downer Landing is one of the most interesting and beautiful spots that is reached by the boats of the Boston and Hingham

Steamboat line. Few people who were familiar with the harbor and surroundings ten years ago would believe that Crow Point could have been transformed into such a charming place as we now find here. Melville Garden, within easy access of the steamboat landing, is the principal centre of attraction for those visiting the place for a day's pleasure. Summer-houses, swings, beautiful shade-trees, a fine collection of monkeys, and above all the most magnificent dancing pavilion that can be found anywhere along the beach, are here for the pleasure of the visitor. A clam-bake, that well rivals Rocky Point, is also to be had here, three times a day. The attractions offered for an evening's entertainment are not a particle behind those of the day, as the whole scene, beautiful in the sunlight, is rendered doubly so by the dazzling rays of the brilliant electric lights recently erected. The numerous diversities which are made in the colors and distribution of these lights have a most gorgeous effect, and immediately suggest to the spectator the richest and most elegant transformation scene. Indeed, it would be very easy to imagine one's self in the enchanted halls of Aladdin. The sweet strains of music wafted to the ear serve to enhance the delusion; but the merry peals of laughter, which come from the dancers at the pavillion, remind the dreamer that he is at Downer Landing, and not in Arabia.

Mr. J. D. Scudder, the manager of these gardens, deserves the heartiest thanks and good-will of the excursionists, for it is to him they are indebted for the pleasures they enjoy when passing a day and evening in these most picturesque gardens.

The hotel, — Rose Standish House, — also under the superintendency of Mr. J. D. Scudder, but directly managed by Mr. F. C. Safford, is another great attraction to the visitor; not so much so to the transient excursionist as to those seeking a few weeks' rest and recreation at the seashore. The house is tested to its utmost capacity every season by some of our first families The high social standing of its proprietor, Mr. Samuel Downer, as well as that of Mr. Scudder, proves a great incentive towards attracting the *élite* of society.

HINGHAM

Is noted for its beautiful scenery and pleasant drives; also for its old meeting-house (erected in 1681), the oldest occupied house of worship in the United States. In the rural cemetery, n the rear of the meeting-house, rest the remains of the late

John A. Andrew, the "War Governor" of Massachusetts, over whose ashes an elegant marble statue has been erected. The last resting-place of Maj.-Gen. Benjamin Lincoln, of Revolutionary fame, is also in the same cemetery.

The only hotel in Hingham is the CUSHING HOUSE, located in the village, convenient to cars as well as boats.

RAGGED ISLAND.

THE RETURN HOME.

The return home is one of the sweetest and most delightful portions of the day's recreation. Seated on deck with your

DOWNER LANDING.

MAGNIFICENT DANCING PAVILION,

Summer Houses, Swings, etc.

CLAM BAKES

THREE TIMES A DAY.

ELECTRIC LIGHTS.

Edmands's Band.

*Those seeking a few weeks' Recreation can
find it at the*

ROSE STANDISH HOUSE.

friends, you remember the many pleasures of the day and rest yourself as you look again upon the surroundings. You may instinctively say with one who has lived in Boston during his life, and who has every season visited these scenes: —

> I have loved thee, Ocean! and my joy
> Of youthful sports was on thy breast to be
> Borne, like thy bubbles, inward; from a boy
> I wantoned with thy breakers, — they to me were a delight;
> For I was, as it were, a child of thee.

Who has not gazed at the noble ships that pass by on the outward or inward trip, with feelings of awe and admiration? See how nobly it comes upon you with all sail set, speeding on its eventful journey; the great white wings, filled with the wind, go rushing by and are now out of sight. Then there are the yachts and pleasure boats, filled with happy voyagers who greet you with cheers or waving handkerchiefs; all become friends, and the joys of each are one in common. If one desires to take the late boat home, the enjoyment is frequently enhanced by a moonlight sail. When the moon is up, though it may not be night, and the sunset divides the sky with her, a sea of glory streams along the shore and waters, heaven is free from clouds; but all the colors seem to be melted to one vast Iris of the West, where day joins past eternity. The face of heaven comes down upon the waters, and all its hues from the rich sunset, and with the moon lend their magical diffusion of varied lights. Watch the change as you sail along; a paler shadow strews its mantle over the deep; parting day dies like a dolphin, whom each pang imbues

> " With a new color as it gasps away,
> The last still loveliest, till — 'tis gone, and all is gray."

There are the lights in the buildings as you near the city, and the reflections of the moon on the "gilded dome," which, in contrast with the shadows of the buildings, form a truly marvellous picture, and complete the round of pleasure and sights enjoyed and seen throughout the day.

When you reach the pier, you immediately feel the change of air, and the fresh, vigorous atmosphere of the ocean is left behind, and you inhale the close odors of the sultry streets. 'Tis then one feels that they would like to live down the harbor during the summer; and rest assured the first spare day that

comes along finds many of the excursionists embracing the opportunity to again enjoy the wholesome, health-giving recreation that is afforded by the Boston and Hingham Steamboat
Company.

HOW TO REACH ROWE'S WHARF.

Out-of-town parties, who arrive at the several railroad depots,
wishing to reach the boats, by taking any Metropolitan car
(some one line of which company passes every depot), can do
so, WITHOUT EXTRA CHARGE, by telling the conductor that
they want to go to ROWE'S WHARF, and he will see that they
are transferred to the right car.

THE PEOPLE'S LINE OF COACHES to Inman Street, Cam-
bridgeport, from the head of Summer Street, Boston, every
eight minutes, from 7 A. M. to 8 P. M.; also barges from Bowdoin and Haymarket Squares, connect with every boat that
leaves ROWE'S WHARF.

HACK FARES.

Parties wishing to employ a carriage to convey them from
ROWE'S WHARF to any portion of the city would do well to
consult the following tariff; viz. : —

For one adult passenger from one place to another within the
city proper (except as hereinafter provided) ; or from one place
to another within the limits of East Boston; or from one
place to another within the limits of South Boston ; or from
one place to another within the limits of Boston Highlands
(formerly Roxbury), the fare shall be 50 cents, and for every
additional adult passenger, 50 cents.

For one adult passenger from any place (within the city
proper) south of Dover Street and west of Berkeley Street to
any place north of State, Court, and Cambridge Streets, or
from any place north of State, Court, and Cambridge Streets
to any place south of Dover Street and west of Berkeley Street,
the fare shall be $1.00, and for two or more passengers, 50 cents
each.

For children under four years of age, with an adult, no
charge shall be made. For children between four and twelve
years of age, when accompanied by an adult, 25 cents each. .

Between the hours of 12 o'clock at night and 6 o'clock in the
morning, for one adult passenger, the fare shall be double the
amount allowed in the preceding sections, and 50 cents for
every additional adult.

BOSTON HIGHLANDS.

For one adult passenger from any place in the city proper north of Essex and Boylston Streets to any place in the Boston Highlands, or from any place in the Boston Highlands to any place in the city proper north of Essex and Boylston Streets, the fare shall be $2.50; for two passengers, $1.25 each; for three passengers, $1.00 each; for four passengers, 75 cents each.

For one adult passenger from any place in the city proper south of Essex and Boylston Streets and north of Dover and Berkeley Streets to any place in the Boston Highlands, or from any place in the Boston Highlands to any place in the city proper south of Essex and Boylston Streets and north of Dover and Berkeley Streets, the fare shall be $2 00; for two passengers, $1 00 each; for three passengers, 75 cents each; for four passengers, 62 and 1-2 cents each.

For one adult passenger from any place in the city proper south of Dover and Berkeley Streets to any place in the Boston Highlands, or from any place in the Boston Highlands to any place in the city proper south of Dover and Berkeley Streets, the fare shall be $1.25; for two passengers, 75 cents each; for three or more passengers, 50 cents each.

BOSTON HOTELS.

ON AMERICAN PLAN.

Revere House, Bowdoin Square; Hampton Hotel, Haymarket Square; Brunswick Hotel, Boylston Street; Tremont House, Tremont Street; Quincy House, Brattle Street; American House, Hanover Street; St. James Hotel, Franklin Square; Creighton House, Tremont Street; Commonwealth Hotel, Washington Street; Adams House and E. P., Washington Street; Evans House, Tremont Street; United States Hotel, Beach Street; Metropolitan Hotel, 1166 Washington Street.

ON EUROPEAN PLAN.

Parker House, School Street; International Hotel, Washington Street; Young's Hotel, Washington Street; Carlton House, Hanover Street; Crawford House, Scollay's Square; Bell's Hotel, Court Square.

BOATS LEAVE BOSTON,

From Rowe's Wharf, 340 Atlantic Avenue,

Junction with Broad and High Streets.

WEEK DAYS.

From Boston to Nantasket Beach at 5.45, 9.30, *10.30, *11.30, A. M.; *12.30, 2.20, 3.35, 5.00, 6.10, 7.10, and †9.30, P. M.

From Boston to Hull at 5.45, 9.30, 11.30, A. M.; 12.30, 2.30, 3.30, 5.30, 6.10, and 6.30, P. M.

From Boston to Downer Landing at 5 45, 9.15, 10.30, 11.30, A. M.; 12.30, 2.30, 3.30, 5 30, 6.30, 7.45, and †‡9.30, P. M.

From Boston to Hingham at 9.15, A. M.; 2.30 and 5.30, P. M.

RETURNING.

From Nantasket Beach to Boston at 7.00, 8.00, *9 30, 11.00, A. M.; *12.00, M.; *1.00, 2.00, 3.30, 4.50, 6.00, 8.30, and 9.30, P. M.

From Hull to Boston at 7.20, 7.50, 10.00, 10.50, A. M,; 1.30, 4.15, 5.30, and 9.45, P. M.

From Downer Landing to Boston at 7.00, 7.35, 9.45, 10.35, A. M.; 12.15, 1.15, 4.00, 5.15, 6.45, ‡8.30, and ‡9.30, P. M.

From Hingham to Boston at 7.30, 10 30, A. M.; 3.40 and 6.30, P. M.

SUNDAY BOATS.

From Boston to Nantasket Beach at 9.30, 10.30, 11.30, A. M.; 12.30, 1.30, 2.30, 3.30, 4.15, 5.15, 6.15, 7.15, and 9.00, P. M.

From Boston to Hull at 10.15, A. M.; 1.30, 2.15, 4.45, 5.15, and 7.15, P. M.

From Boston to Downer Landing at 10.15, A. M.; 2.15, 4.45, and 7.15, P. M.

RETURNING.

From Nantasket Beach to Boston at 11 00, A M.; 12.00, M.; 1.00, 2.00, 3.00, 4.00, 5.00, 6.00, 7.00, 8.00, and 9.30, P. M.

From Hull to Boston at 12.15, 3.45, 4.15, 6.15, 8.15, and 9.45, P. M.

From Downer Landing to Boston at 12.00, M.; 3.30, 6.00, and 9.30, P. M.

* Via Downer Landing. † Saturdays excepted. ‡ Mondays excepted.

25c.   Fare Each Way.   25c.

Excursion Tickets, including admission to Melville Garden at Downer Landing, 60c., for sale at ticket office in Boston, except on Mondays and Holidays.

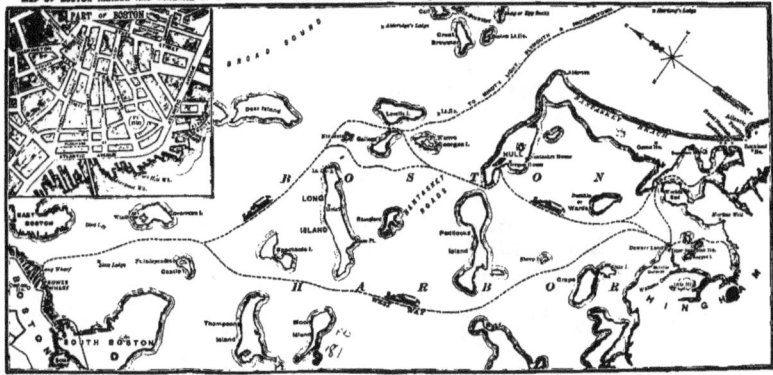

MAP OF BOSTON HARBOR AND DIAGRAM OF STREETS LEADING TO ROWE'S WHARF, the Landing-Place in Boston of the Steamers of the Boston & Hingham Steamboat Company.

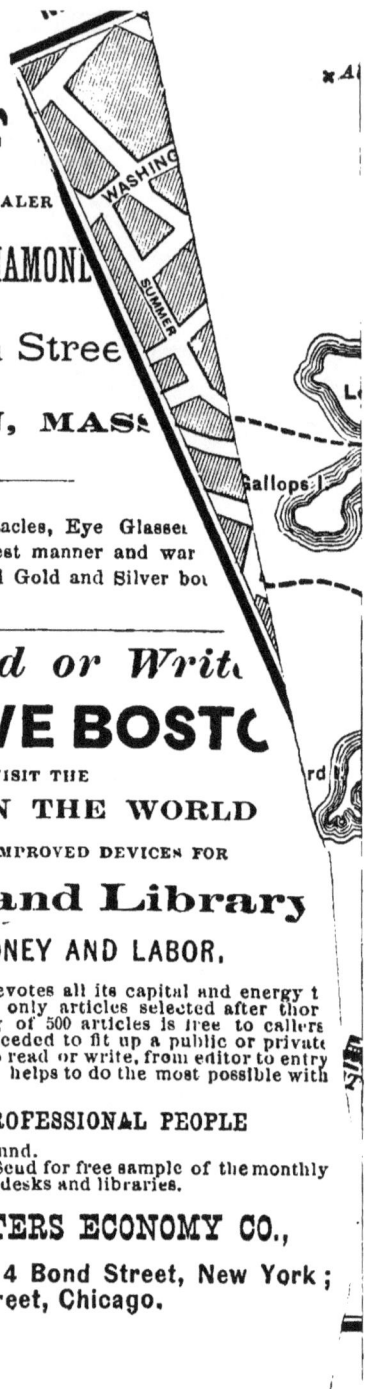

Boston & Hingham Steamboat Co.'s Guid

and

SEASHORE RESORTS.

BIRD'S-EYE VIEW OF
MELVILLE GARDEN, DOWNER LANDING.